Arthur Jacob

Practical Designing of Retaining Walls

Arthur Jacob

Practical Designing of Retaining Walls

ISBN/EAN: 9783337011697

Printed in Europe, USA, Canada, Australia, Japan

Cover: Foto ©Lupo / pixelio.de

More available books at **www.hansebooks.com**

PRACTICAL DESIGNING

OF

RETAINING WALLS.

BY

ARTHUR JACOB, A. B.,

ASSOCIATE OF THE INSTITUTE OF CIVIL ENGINEERS ; LATE EX-
ECUTIVE ENGINEER, H. M. BOMBAY SERVICE.

NEW YORK:

D. VAN NOSTRAND, PUBLISHER,

23 MURRAY AND 27 WARREN STREET.

1873.

PRACTICAL DESIGNING

RETAINING WALLS.*

———•••———

In designing masonry works there is
hardly any subject that presents itself more
frequently than the retaining or revetment wall; and in some form or other it is
found to enter into almost every design.
To the military engineer no less than to
his civil brother is the subject one of importance and interest, forming as the revetment wall does, for the most part, a
component element in works of defence.
To military engineers in truth is due some
of the most valuable information that

---

* Practical Designing of Retaining Walls, by Arthur Jacob,
A. B.

civil engineers possess regarding the theory of earth pressure, and although further considerations are involved in designing revetments for military works than the mere support of earthwork, there is still to be derived from the experiments and researches of military men information of much value to civil engineers. The subject is one that has received the fullest and most able treatment at the hands of mathematicians, and solutions for every case that could possibly occur in practice are to be found in our text-books. But the mathematical investigations of this and many other questions of common occurrence in practice, unquestionably valuable as they are, in determining the principle involved, and establishing final rules applicable to practice, are, it is believed, but rarely resorted to by practical engineers. Even when such examples have to be dealt with by those sufficiently acquainted with the mathematical mode of proceeding, they are generally solved without hesitation by some empirical rule, derived from experience. Such a method may, and doubtless occasionally does, lead to accident from weakness, and not unfrequently to clumsy waste of material and consequent expense. But it is not clear that less of failure or clumsiness would result if every retaining wall were calculated with mathematical precision, for in truth the data involved are so variable and imperfect, and the disturbing causes are of such a character as to neutralize to a great extent the accuracy of the investigation. With certain specific data theoretical accuracy can always be attained; but the engineer as a rule knows nothing with absolute certainty either of the weight of the earth he has to sustain in position, or of the masonry that he intends to adopt in doing so. These and other data he must assume before he enters on his calculations; and though there is not in these, as in many other investigations, any necessity to attempt an extreme degree of refinement, which would be inapplicable for every-day practice, yet there can be no good excuse for dealing with the matter by hap-hazard and guess-work.

It is not proposed now to regard with

more than a cursory glance the principles involved in determining the strength of walls to support earthwork. Such simple rules will be given, as it is hoped will serve—due regard being had to the peculiarities of each particular case—to guide the less experienced in designing works of this class. The empirical mode of dealing with the question is clumsy and unscientific, whilst the formulæ usually given are so complicated as to render their application to practice out of the question.

SPECIFIC CAUSES OF FAILURE.

It must not be presumed that the failure and destruction of a retaining wall is necessarily due to the wall being of itself insufficiently strong. It may be quite heavy enough to resist the pressure of a bank, if due regard be had to the mode of forming the earthwork, and to drainage; but if these points be not fairly considered and observed at first, a retaining wall of quite sufficient thickness will probably give way sooner or later. As much care should in fact be devoted to the method of backing up and draining a wall, as to the calculation of its section; for indeed if these matters be disregarded, no retaining wall, properly so called, can be implicitly relied upon to stand. With the exception of one particular case, which will be noticed hereafter, walls are designed on the assumption that they are to support a dry material—or one, at any rate, not permeated by water—and that the material is to be deposited in such a manner as to have no predisposition to slide against the wall. It is, of course, also presumed that the wall shall be of fair workmanship and materials, and where these points cannot be relied upon, as is sometimes the case, especially in foreign works, some allowance should be made in the dimensions of the wall. It has not unfrequently happened that a retaining wall will have stood for a considerable number of years without showing any appearance of yielding, and yet will give way suddenly and completely, without apparent cause. Such failures can generally be accounted for by the fact of the wall not being designed to resist a

maximum pressure, and never having been
tried fully till the time of its destruction.
Much apparent anomaly is observed in the
way that retaining walls are found to fulfil
the purpose for which they are designed:
for whilst some will yield, others of less
dimensions will continue to stand; such
apparent inconsistency giving occasion for
ingenious theories, most of them entirely
unsupported by fact or experience. The
truth is, that imperfect drainage, defective
foundations, or rotten work will account for
almost every conceivable case of failure.

FIRST CASE—HYDROSTATIC PRESSURE.

The first and simplest case of a retaining
wall to be considered is that in which the
pressure of water has to be counteracted;
not indeed that the question in such a form
belongs strictly to the subject under notice;
but it nevertheless becomes absolutely the
method of determining the strength of walls
for certain positions. It not unfrequently
happens, as in some hydraulic works or
with the wing-walls of aqueducts, that the
infiltration from behind, which is not al-

ways avoidable, may produce such a pressure as no retaining wall properly so called could be expected to bear. With this view the engineer's limit of safety will be attained when the structure is designed to sustain the full hydrostatic pressure. The pressure of water upon any plane surface immersed is known to be *equal to the area of that surface, multiplied by the depth of its centre of gravity below the level of the water, and by the weight of a unit of water.* Generally speaking the unit adopted in calculation is a foot; and the unit of water being taken at a cubic foot, weighing 62.5 lbs., the resulting product, from the multiplication of the three quantities, will give the *pressure* in pounds on the surface immersed. Let it be supposed for simplicity that water to the depth of 10 ft. has to be sustained by a vertical rectangular wall. It is usual to take but 1 ft. length of the wall for the calculation, though it will not affect the result whether 1 ft. or 100 ft. be the length assumed. We then have the surface under pressure = 10 sq. ft., the depth of the centre of gravity = 5 ft., and the weight of a cubic foot of water = 62.5 lbs. ; the product of which quantities give us 3,125 lbs., the *pressure* on 1 ft. length of the wall. But this pressure is not the whole of the force that the wall has to resist; the leverage that it exerts must also be taken into account. In the example under consideration—namely, that of a vertical plane, with one of its sides coinciding with the surface of the water, as in Fig. 1.—the whole of the pressure is so distributed as to be equal to a single force acting at a point one-third of the depth from the bottom. Thus the total force to be resisted by the wall is 3,125 × 3.33 = 10,416, which is the *moment* tending to overturn the wall.

MOMENT OF RESISTANCE TO OVERTURNING.

It is evident that a certain weight of wall must be opposed to this overturning force; and as the height of the wall and the length are determined quantities, the thickness alone remains for adjustment. But as a rectangular wall in upsetting is considered to turn upon a single point, F, Fig. 1.—namely, the outer line of the foot

of the wall, there will be a certain amount
of leverage to assist the wall in resisting
the pressure of the water. This leverage is
the horizontal distance of the centre of
gravity of the wall from the turning point
F, and when the structure is rectangular
and vertical, it is equal to half the thickness. *The amount of the wall's resistance
will then be equal to the number of cubic
feet in one foot of its length, multiplied by
the weight of a single cubic foot of masonry,
and by half the thickness of the wall.*
Taking w = the weight of a cubic foot of
water = 62.5 lbs., w^1 = the weight of a
cubic foot of masonry, say 112 lbs.; x =
thickness of the wall, and h = the height;
the condition of simple stability will be ful-
filled when

$$w^1 \times h \times x \times \frac{x}{2} = w \times h \times \frac{h}{2} \times \frac{h}{3} \quad (1)$$

$$\frac{w^1 h x^2}{2} = \frac{w h^3}{6}$$

and solving for x we get

$$x = \sqrt{\frac{w^3 h^2}{3 w^1}} \quad \cdot \quad \cdot \quad \cdot \quad (2)$$

The thickness of the wall = 4 ft. 4 in.

EXAMPLE.

A simple example has been selected for illustration, but of course a rectangular section of wall would not be found generally applicable in practice, nor would it be expedient to limit the dimensions of a retaining wall of whatever kind to the minimum that would sustain the pressure; some margin of safety must therefore be allowed, to cover inferior work and materials. It is true that no account has been taken of cohesion, which, if the wall be founded on rock or concrete, may be assumed to add to its stability about 7,000 lbs. for every square foot of base. In addition to this, practice seems to indicate an increase on the calculated thickness, and in the example the mean width might be augmented to 5 ft., the stability being further increased by altering the section from a rectangle to a battering wall with offsets at the back.

A good general rule for the dimensions of a wall designed to support water or earth in a semi-fluid condition will be—

Top breadth = 0.3
Middle do. = 0.5
Bottom do. = 0.7

The height being represented by unity.

Proceeding to the consideration of walls for the support of dry earth, it will be found that the question is one that will in general require the engineer to exercise his judgment, to determine what angle of repose he will base his calculation upon. The natural slopes assumed by earths of different tenacity are so various, that an average figure cannot be adopted with safety; the calculation of pressure from earth, in fact, depends essentially on this point, and a disregard of it will lead to very doubtful results. The following are a few of the slopes assumed by different materials, but it is probable that the engineer's judgment will be of more service than any table in deciding the angle of repose. The examination of a district in which works are intended to be carried out will always suffice to satisfy the designer of the nature of the material that he is dealing with, and may enable him to proportion his works very nearly to the requirements of safety and economy :—

	Angle of repose.		Slope.	
Dry sand, clay and mixed earth	From 37°	..	1.33	to 1*
	to 21°	..	2.62	to 1
Damp clay..............	45°	..	1	to 1
Wet clay	From 17°	..	8.23	to 1
	to 14°	..	4	to 1
Shingles and gravel......	From 48°	..	0.9	to 1
	to 35°	..	1.43	to 1
Peat........:..........	From 45°	..	1	to 1
	to 14°	..	4	to 1

To which might be added as a special feature London clay; it appears under the influence of weather to be exceedingly unstable, slipping away to almost any angle of repose.

THEORY OF EARTH PRESSURE.

It has been ascertained by M. Prony that when a vertical wall sustains the pressure of a bank of earth the top of which is horizontal, the maximum horizontal pressure to which the wall can be subjected will be reached when the plane of fracture of the earth bisects the angle that would be formed were the earth to slope from the

---

* Rankine's "Manual of Civil Engineering."

foot of the wall backwards at the natural inclination. This fact is somewhat striking, for it would appear at first sight, and was for long assumed, that the angle of fracture ought to coincide with the natural slope of the earth; such is, however, not really the case. If we suppose the angle made between the sloping plane and the vertical to be bisected, the prism of earth enclosed between the bisecting plane and the wall will represent the mass, the pressure of which has to bé resisted; and this being the *maximum* pressure that a horizontal topped bank is capable of exerting, it is usually the point to be determined.

Referring to Fig. 1, the principle of earth pressure will readily be understood. Supposing the plane of rupture to bisect the angle c—which will be the case when the pressure is a maximum—the prism cut off will be the whole weight that the wall will have to sustain. Taking this prism for a single unit of length or thickness, the superficial area will represent the cubic contents. But the area of the triangle, taking h as the height of the wall, will be

$$\frac{h^2 \tan. \frac{1}{2} c}{2}$$

c being the angle contained between the natural slope of the earth and the back of the wall. It is only necessary to multiply this value by w, the weight of a cubic foot of the bank, to get the total weight of the prism.

FIG. 1.

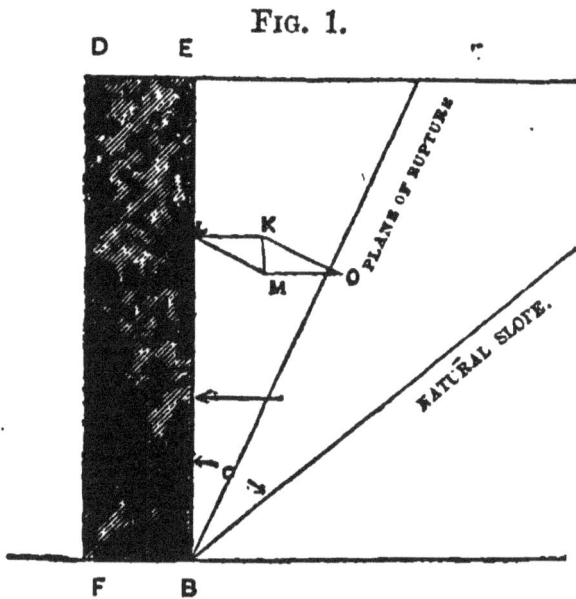

This prism of earth is then like any other body resting upon an inclined plane; which in this case is the plane of rupture. It is sustained in position by the wall on one side and by the fixed portion of the bank

on the other; and may be regarded as a
solid mass of material without motion
amongst its parts. The line K M represents the direction of the force of gravity,
and the lines K L and K O the pressures
exercised against the wall, and the force of
the bank respectively. These pressures
produce a certain amount of friction against
the wall and the bank, but, as the friction
against the wall does not materially affect
the question, the friction of the bank alone
is considered, and taken into account in
arriving at the following formula which
applies to the case of a vertical wall supporting a bank with a horizontal-topped
bank :—

$$P = \frac{w\,h^2}{2}\ \tan.\ ^2\ \tfrac{1}{2}\,c\ .\ \ .\ \ (3)$$

Having calculated the *pressure* of the earth,
the next step will be to determine its *moment* to overturn the wall, and this can be
ascertained, as in the case of water, by
multiplying the pressure by one-third of
the wall's height. This having been determined the next consideration will be,
what weight of wall will suffice to sustain

it; and the method of arriving at this is similar for the most part to that adopted for water. Taking, as above, the moment of the wall to resist the pressure, the following equation will represent the conditions of stability :—

$$\frac{w' h x^2}{2} = \frac{w h^2}{2} \tan.^2 \tfrac{1}{2} c \frac{h}{3}$$

And solving for x, the thickness of the wall, we have—

$$x = \sqrt{\frac{w h^2 \tan.^2 \tfrac{1}{2} c}{3 w'}} \quad . \quad (4)$$

If the weight of a cubic foot of earth be taken equal to a cubic foot of the wall, the value will be—

$$x = \sqrt{\frac{h^2 \tan.^2 \tfrac{1}{2} c}{3}} \quad . \quad (5)$$

which would give a thickness of 2.69 ft. for a rectangular wall of 10 ft. high supporting a bank of earth, the angle of repose being taken at 40 deg. The average weight of brickwork and ordinary clay will generally be nearly the same; but if great accuracy be desired, and the respective weights of the materials be known, the general formula No. 4 must be used.

The following table gives the weight per cubic foot in pounds avoirdupois of such materials as come under our consideration in solving questions relative to retaining walls:

	Weight of a cubic foot in pounds.
Sand—damp	120
Do. dry	90
Marl	100
Clay	120
Gravel	125
Brick	130
Brickwork	112
Masonry	130
Mortar	110

PARTIAL RETAINING WALL.

Having so far considered the first two cases, namely, those of a wall supporting a horizontal-topped bank of earth in a semi-fluid condition, and also in a state of comparative dryness, the next example that suggests itself to our notice for examination is that of a partial retaining wall, or a wall from the top of which the bank slopes away for a certain height—called the surcharge —either at the natural slope of the earth or at a less inclination. Such mode of construction is of very common occurrence, dwarf walls being frequently adopted on railway works where the cuttings or embankments are of considerable height, and when carefully designed are found to effect a saving of expense, both in construction and in the item of land. In cuttings the walls are carried up to such a height as economy dictates, and the slope is then trimmed back at the proper angle. Similarly with embankments, the walls are so disposed as to cut off the foot of the slope. In either case a little consideration will suffice to show whether the saving of earth and land area will cover the cost of the retaining walls. In military works, as well as civil, the partial revetment is very commonly used, being, indeed, a component part of almost every system of fortification.

The first particular case belonging to this class, though not of the commonest occurrence in civil practice, is when a partial retaining wall supports a bank, the face of which slopes back at an angle less than the natural slope of the earth. As M. Prony's rule, that the plane of rupture bisects the angle between the natural slope of the earth and the back of the wall, only holds good when the surface of the bank is at right angles to the plane of the wall, another mode of determining the angle for the maximum pressure must be resorted to. The simple construction given in the note enables us to arrive at the maximum pressure for a wall at any given batter, with the surcharge above sloping at any inclination. The equation arrived at is the expression for the maximum horizontal pressure:

$$P = \frac{w\,h^2}{2} \times \frac{\tan.\ \theta \tan.^2 (c - \phi)}{\tan.\ \theta - \tan.\ c} \quad . \quad . \ (6)$$

the angle c being that between the back of the wall and the natural slope; $\theta =$ the angle made by the face of the bank with the plane of the wall; and $\phi =$ the angle between the plane of rupture and the back of the wall. The value for $c - \phi$ will be found in the note. Taking, for example, a vertical wall of 10 ft. high, supporting a bank that slopes back at an inclination of 20 deg. with the horizon, the natural slope being 40 deg., the value of tan. ($c. - \phi$) will be

.4610; inserting this value and working out the equation, we arrive at a pressure of 2,100 lbs. against the back of the wall.

For the case of a revetment sustaining a surcharge the centre of pressure will be, as in the former case, at one-third of the height of the wall, giving a leverage of 3.33 feet. This gives 2,100 lbs. $\times$ 3.33 = 6,993, the moment of the earth tending to overturn the revetment. Equating this value to the moment of the wall, taking the cube foot of brickwork at 112 lbs., the same weight as the earth, and solving for x the thickness, we find it to be 3.53 ft.

DEFINITE SURCHARGE.

The next case to be considered is one of much more frequent occurrence in practice than that just mentioned; it is a partial retaining wall supporting a surcharge of earth, sloping away at the *natural* inclination, and terminating in a horizontal plane above. Cuttings and embankments partly supported by masonry works furnish familiar examples of this, which is denominated the " definite surcharge." The most convenient

method of determing the thickness of wall
in this instance will be to consider, first, the
conditions of stability for an infinitely long
siope, which, however, can only have a

Fig. 2.

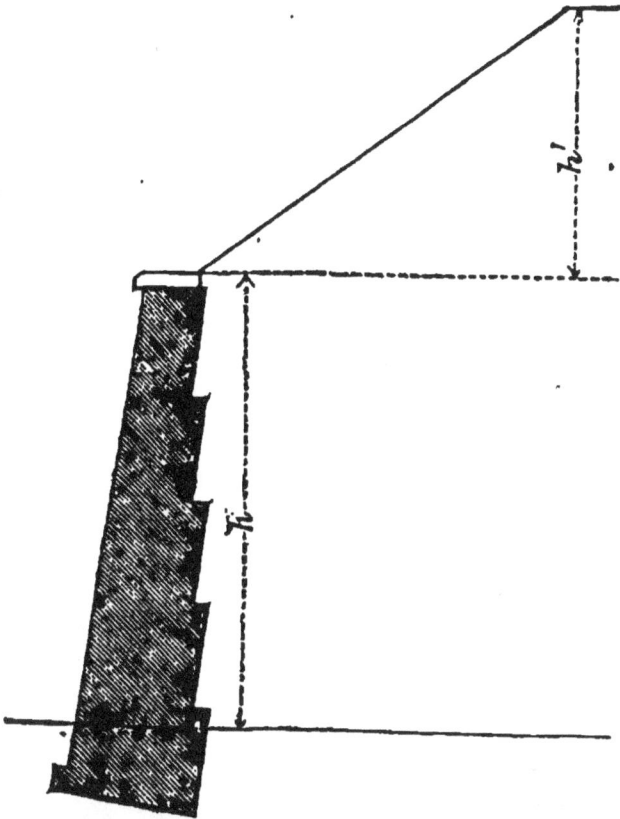

theoretical existence; and having arrived
at the thickness of wall necessary to support
such a bank, a simple reduction will give

the thickness required when the length of slope is limited.

It has been mentioned that when a vertical wall sustains a bank with a horizontal top, the plane of rupture for the maximum pressure is found to bisect the angle between the natural slope and the vertical. It is also an ascertained fact, that as the angle of the surcharge increases, the angle ϕ, or that between the plane of rupture and the back of the wall, also increases; until the face of the bank slopes at the natural inclination of the earth, and then the plane of rupture becomes parallel to it. From this it would appear that when the slope is infinitely long—a condition that could not exist in practice—the pressure will also be infinitely great; but such is not really the case. The ratio of the pressure of a bank, whatever its inclination, to the pressure exerted by an embankment level with the top of the wall can never exceed 4 : 1. The formula, then, for finding the maximum horizontal pressure exerted by an infinitely long slope against a vertical wall will be—

$$P = \frac{w\,h^2}{2}\,\text{sin.}^2\,c \quad . \quad . \quad (7)$$

the notation being exactly the same as in the other cases. If we work this pressure out, using the same values for w, h, and c, as taken above, we shall find P=3,281 lbs.

Now for the leverage: we have, as in every other case, simply to divide the height of the wall by 3, which in our example gives 3.33 and the moment to overturn the wall $= 3,281 \times 3.33 = 10,925$. Proceeding in the same manner as before, the width of a wall of brick to counterbalance an infinitely high bank sloping at the natural inclination, will be found to be 4.43 ft.

When the surcharge is very high as compared to the height of the wall, no reduction of the thickness will be necessary, for practically the slope may be considered infinite; but when the bank does not overtop the wall by a great height it will be well to apply the following formula to ascertain the corrected thickness. Let $h=$ height of wall=10 ft., $h'=$height of surcharge above the wall, which we shall take

at 20 ft., t=thickness of wall to support a
horizontal bank, as found in the first case
=2.69 ft., T=the thickness of a wall for
20 ft. surcharge, t'=thickness for indefinite
slope as found=4.43. Working this out
the thickness is found to be 408 ft.

$$T = \frac{h\,t + 2\,h'\,t'}{h + 2\,h'} \quad \cdot \quad \cdot \quad \cdot \quad (8)$$

So far we have considered the cases of
more usual occurrence in practice, namely
those in which the back of the wall is vertical or stepped, which is practically the same
thing. For the calculation of leaning walls
the reader is referred to the general formulas (A) and (B) given in the note; from the
latter formula the horizontal resistance of
any bank, supported by a wall at any angle
of inclination, can be ascertained, and the
leverage being in every case taken at one-
third of the height of the wall, there will be
no difficulty in designing a wall of such a
section as will resist the pressure of the
bank effectually.

The point to be kept in view is the moment of the wall, and this must be made

to *exceed* the maximum overturning force of the embankment. It will not suffice to *equalize* the moment of the earth's force to the resistance of the wall, as has been done

FIG. 3.

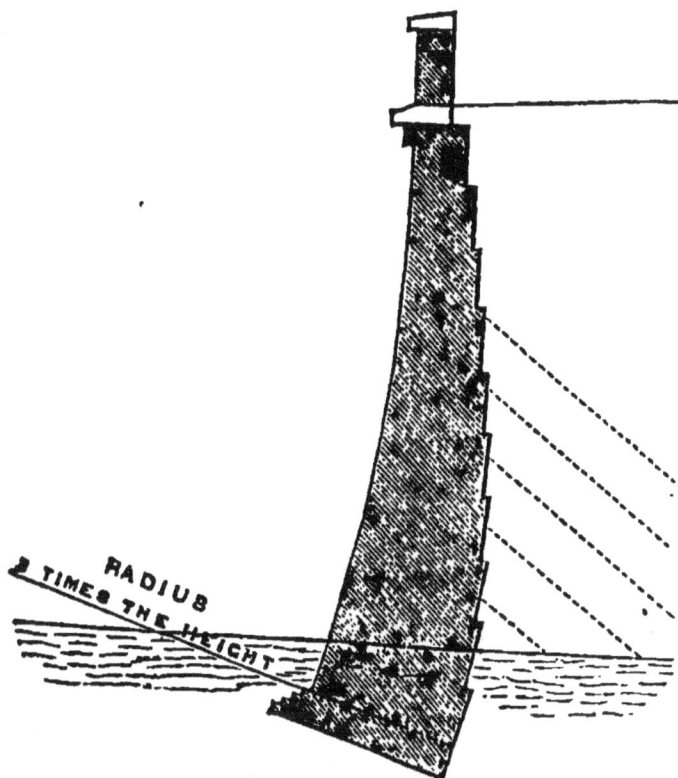

in the examples above; a certain excess of resistance will be necessary, and this can easily be attained by giving the wall a batter, or else sloping it back so as to throw the centre of gravity of the mass as far back as possible, in a horizontal direction from the outer line of the foot of the wall. The line of the centre of gravity must not, however, be allowed to fall inside the base of the wall, otherwise the stability of the structure will become dependent on the support of the bank, and will have a tendency to slide away from its position.

It has been stated, and taken for granted, that banks of earth, when they destroy retaining walls, do so by turning them over; this is, however, not invariably the case. It has occasionally happened that walls have been moved bodily forward, sliding on their base. Such an occurrence is certainly accidental, and is probably the result of the wall having been founded on an unstable material, perhaps on an inclined bed of moist and uncertain soil. Walls have also given way in rare instances by the upper courses of the structure yielding to pressure, breaking off and falling over; a contingency that is probably due to the upper part of the bank becoming suddenly charged with water, and exercising an undue pres-

sure on the wall before there is time for the
water to drain away. These must be rè-
garded as rare contingencies, arising out of
some defect of the foundations, or backing;
and cannot affect the consideration of the
wall's stability generally. The theory of
the wall being turned over on its base pro-
vides for the greatest trial to which the
structure can be subjected, or, in other words,
the wall would as a general rule give way
under a much less pressure by falling over,
than would be required to overcome friction, and move the wall forward in its en-
tire state; if therefore the structure is considered as having to withstand the overturning force, it will always be strong
enough to resist being pushed forward.

RETAINING WALL WITH CURVED BATTER.

A form of retaining wall commonly met
with in practice, especially in brickwork
structures, is that with a curved batter,
stepped in offsets at the back. The curve
usually adopted is the arc of a circle, the
radius of which is from 2½ to 3 times the
wall's height; and the centre of the curve

is as a rule in the same horizontal plane as
the top of the wall. In such structures the
courses are made to radiate from the centre,
and the result is that the joints of the
brickwork at the back are thicker than is
either necessary or advisable. When the
radius of curvature is large, the increase of
thickness is inconsiderable, but it becomes
decidedly an objection when . the curve is a
short one; for the thickness of the wall will
not become reduced in the same proportion
as the height or as the radius of curvature.
The dimensions of a wall of this kind may
be determined with sufficient accuracy, by
first considering it as a leaning wall at a
given slope, and using the general formula
(6), and in this manner a very close ap-
proximation to the thickness may be arrived
at. There are, it is true, specific formulæ
given by some authors for determining the
thickness of curved walls, but they are too
complex for application in practice. The
effect of the curvature will be to add to the
wall's stability by bringing the centre of
gravity farther in towards the bank, and
this, indeed, is the only advantage that the

curved form possesses; it is difficult to construct, and consequently expensive; for the saving of material, if any, is very trifling. In architectural effect it certainly has no advantage over the wall with a straight batter, for the simple reason that it does not convey the same idea of strength. If the curved wall is supposed to derive any additional stability from its curvature, on the principle of the arch, as some have fancied, it must be recollected that an arch with but one abutment is a very unstable kind of structure, and such kind is simply what the curved retaining wall is. Quays and river walls may, indeed, be designed of a curved form with advantage, for such will allow of ships coming closer to the brink, than they could were the wall a straight one. And sea walls, also, are not unfrequently built of a curved section on the face, this form being under certain circumstances better adapted than a straight wall to resist the force of waves.

In situations where a retaining wall has but one purpose to fulfil—that of supporting a bank of earth—it is usual to give the base of the wall a certain amount of inclination to the horizontal, the slope being perpendicular to the batter of the face; or if the wall have a curved batter, the plane of the base will simply radiate from the centre of curvature. Such mode of construction is calculated to increase the frictional stability, for it brings the thrust of the earth from behind more nearly perpendicular to the bearing surface.

COUNTERFORTS.

Counterforts are frequently constructed at the back of retaining walls, and, although generally approved of, appear to be a somewhat doubtful mode of distributing material. Mr. Hosking, in a paper read before the Institute of Civil Engineers, deprecates their use and, with some reason, advocates the use of ribs or arches from wall to wall. These ribs seem to have been suggested by the cast-iron beams used to support the falling walls on the London and North Western Railway between Euston Station and Primrose Hill. Mr. Hosking proposes that his arches of brick

should pass completely over the road, and
that they should consist on plan of a pair
of flat arches placed back to back. Such
an arrangement would doubtless prove
effective, and the expediency of adopting it
would evidently be determined by the cost
of the work and the value of land adjoin-
ing—a mode of construction in common
use in metropolitan works, and in other
situations where land is very valuable, is
that shown in Fig. 4. It consists of a series.

FIG. 4.

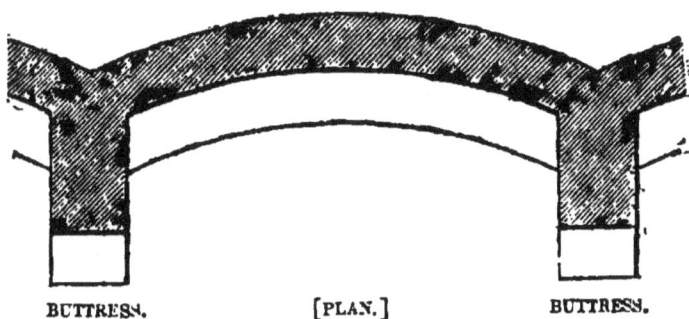

BUTTRESS. [PLAN.] BUTTRESS.

of buttresses and inverts, the convexity of
which latter is opposed to the thrust of the
backing. Such a distribution of material
is most suitable in situations where the
projection of the buttresses is not found
inconvenient. In quay and river walls it

would not answer of course to have any such projection, as the near approach of ships and boats is an essential consideration.

The distribution of the material in the form of counterforts is attended with a slight saving, and where buttresses would be inadmissible on account of their encroaching on the roadway, counterforts may be adopted. They have at least one use, that they oppose more friction to the earth than a plain wall, and, being easy of construction, are productive of but little additional expense. In order to ascertain what additional mean thickness a wall derives from the counterforts, it is only necessary to *multiply the length of the counterfort by its mean width, and divide the product by the distance from centre to centre* of two counterforts. The form and dimensions of counterforts vary with circumstances, the narrow and deep disposition of the material being probably the best as a general rule. The late Lieutenant Hope, of the Royal Engineers, conducted some interesting experiments on the stability of retaining walls generally, and arrived at the conclusion that a thin wall, with frequent thin counterforts, was the best arrangement of the material.

Two points of importance relative to counterforts demand particular attention—the first, that they should be built simultaneously with the wall; and the second, that the wall should be well bonded into the counterforts, otherwise they detract from the wall's strength, instead of augmenting it. It is evident that without some special system of bond, counterforts reducing the thickness of the wall, as they are generally understood to do, must prove detrimental rather than advantageous; but if plenty of hoop iron be used, which is not usually the case, counterforts may be made to contribute in a very considerable degree to the stability of the wall. In fact, quite as much as buttresses.

MODE OF BACKING AND DRAINAGE.

That accidents frequently occur from due care not being exercised in the mode of backing-up retaining walls is undoubted,

and indeed to this cause alone the majority
of failures is attributable; not, as is frequently supposed, to the insufficient section
of the wall. The drainage of masses of
earth sustained by walls, is a matter that
can only be disregarded with risk of ill
consequences. It is a difficult thing to prevent surface water from finding its way
into earth-work, and therefore the simplest
method of dealing with it will be to provide
efficient means for its escape. To this end
holes or *weepers* should be left in the wall
at different levels, to relieve it from pressure
from behind; and in order to admit the
surface water to these points of escape, it
will be advisable to back up the wall with
dry stone, quarry shivers, or whatever else
will admit the free passage of water. If a
wall be backed up in this way by a rough
angular material, it will be relieved of almost all pressure from the earth. Economy
will, however, generally preclude such an
expedient in works of considerable extent,
and then it will be necessary to form the
embankment with great care, adopting every
precaution to prevent the tendency of the

earth to slip in the direction of the wall. It will be evident from the calculation of the pressure exerted by earth, that the less the angle of repose is, the greater will be the pressure on the wall; and, as a matter of course, any means that will tend to increase the angle of repose, will relieve the wall of a certain amount of pressure. Effectual drainage will do much towards this end; but the mode of depositing the earth will also affect the angle of natural slope in a considerable degree. The same earth under different treatment will assume different slopes; if dry, it will fall when tipped—at a low angle, but if damped, and well rammed, will adapt itself to a much higher one. It has even been found that a bank when constructed in such a manner has stood for a considerable time perfectly vertical. The best mode of backing a wall up with earth will then be, to commence depositing at the foot of the wall, and to lay the earth in layers inclining *against* the wall, as shown by the dotted lines in Fig. 3, each layer being well rammed before another is commenced. This will not only consolidate the earth, and prevent any shock that might occur from sudden settlement, but will increase the angle of repose, and give the earth a tendency to slip away from the wall, rather than towards it.

THE LAND TIE.

An expedient for securing retaining walls that is simple and not expensive, is the land tie; it consists of an iron plate, with a rod passing through its centre, the plate being placed vertically in the bank behind the wall, and the end of the rod passed through the wall and secured. The holding power will depend on the area of the plate, and the depth at which it is sunk beneath the surface. But it is evident that, in order to act most effectually, land ties should be attached to the wall at the height of the centre of pressure.

NOTE I.

The following construction, given by Mr. Neville in the "Transactions of the Institute of Civil Engineers, Ireland," vol. i., shows the method of determining the

pressure exerted by a bank, whatever may
be its inclination :

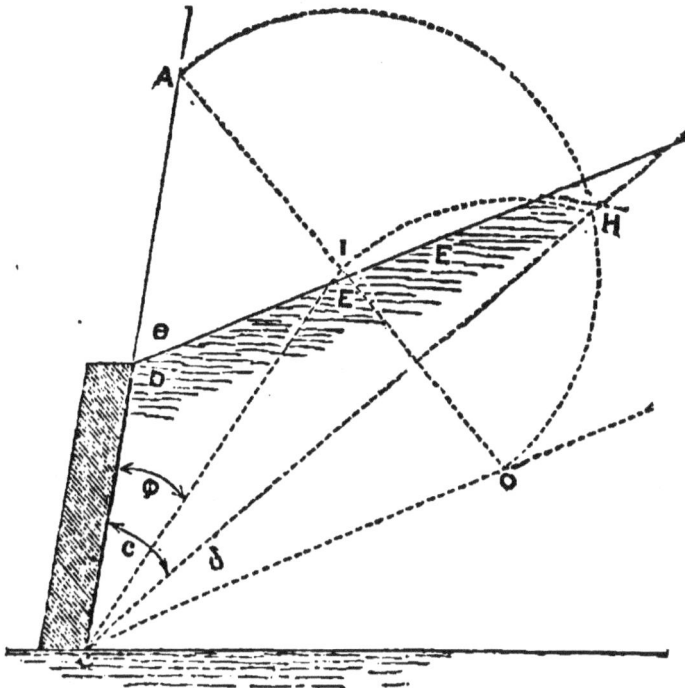

FIG. 5.

Let C D represent the wall; D E the
face of the bank sloping at any angle; and
C H the line of natural slope. Draw any
line perpendicular to the line C H, cutting
the line of the wall produced at A, and also
a line drawn parallel to the face slope at
O. On A O describe a semicircle. From

O, as a centre with the radius O H, describe an arc cutting A O in I : draw I C. The triangle C D F represents the maximum to be resisted. The angle $\delta = \theta - c$. The complement of the angle of repose $= \rho$; and the face C D $= h$

$$\tan. (c-\phi)=(\tan.^2 \delta + \tan. c \tan. \delta) \tfrac{1}{2} - \tan. \delta. \quad \text{(A)}$$

Putting R for the maximum *horizontal* resistance, and w for the weight of a cubic unit of the bank, the resistance of pressure will be

$$R = \frac{w\,h^2}{2}\,\frac{\tan. \theta \tan.^2 (c - \phi)}{\tan. \theta - \tan. c.} \quad \cdot \cdot \quad \text{(B)}$$

in which the value $(c-\phi)$ found above must be substituted. When C D E is a right angle we shall have

$$R = \frac{w\,h^2}{2}\,\tan.^2 \tfrac{1}{2} c \quad \cdot \quad \cdot \quad \text{(C)}$$

the equation given in the first part of this article; and that which holds good when the slope of the bank is at right angles to the face of the wall.

The following tables calculated by Mr. J. H. E. Hart, Executive Engineer of the

Bombay Department of Public Works, are, by his kind permission, appended to this pamphlet, and will be found very convenient for the calculation of Retaining Walls.

Knowing the angle of repose of the earth to be supported, and the relative weights of the masonry of the wall and of the earth per cubic unit, a simple reference to the table will give a coefficient, which multiplied by the height will give the requisite thickness. For example, supposing a *horizontal* topped bank has to be supported by a masonry wall of 10 ft. high, and of twice the specific gravity of the earth, the angle of repose of the latter being 35 deg. Under the fraction $\frac{1}{2}$, and opposite to 35 deg., will be found in Table A the fraction .212, which multiplied by 10, the height of the wall, gives 2.12 ft., the required mean thickness.

TABLE A.

Table of coefficients of h for finding the Thickness of Standard Rectangular Wal's, when the top of the bank is horizontal.

Angle of Repose.	1	1·1	1·2	1·3	1·4	1·5	1·6 (K)	1·7 (K)	1·8 (K)	1·9	2	2·1	2·2	2·3	2·4	2·5
0	.577	.550	.527	.506	.488	.472	.456	.443	.430	.419	.408	.399	.389	.381	.373	.365
30°	.334	.317	.304	.292	.282	.272	.263	.256	.248	.242	.235	.230	.224	.220	.215	.211
31°	.327	.311	.298	.286	.276	.267	.258	.251	.243	.237	.231	.225	.220	.216	.211	.207
32°	.320	.305	.292	.281	.270	.261	.253	.245	.238	.232	.226	.221	.216	.211	.206	.202
33°	.314	.299	.286	.275	.265	.256	.248	.240	.234	.228	.222	.217	.212	.207	.203	.198
34°	.307	.293	.280	.269	.260	.251	.243	.236	.229	.223	.217	.212	.207	.203	.198	.191
35°	.300	.286	.274	.263	.254	.245	.237	.230	.224	.218	.212	.207	.202	.198	.194	.190
36°	.295	.280	.268	.258	.248	.240	.232	.225	.219	.213	.208	.203	.198	.194	.190	.186
37°	.288	.274	.262	.252	.243	.235	.227	.221	.214	.209	.203	.199	.194	.190	.186	.182
38°	.282	.268	.257	.246	.238	.230	.222	.216	.209	.204	.199	.194	.189	.185	.182	.178
39°	.276	.262	.251	.241	.233	.225	.217	.211	.205	.199	.194	.190	.185	.180	.177	.174
40°	.269	.256	.245	.235	.227	.219	.212	.206	.200	.195	.190	.185	.180	.177	.173	.170
41°	.263	.250	.240	.230	.222	.215	.207	.202	.196	.192	.186	.181	.177	.173	.170	.166
42°	.257	.244	.234	.225	.216	.210	.202	.197	.191	.186	.181	.177	.173	.169	.165	.162
43°	.251	.239	.229	.220	.212	.205	.198	.192	.187	.182	.177	.173	.169	.164	.162	.158
44°	.245	.233	.223	.214	.206	.200	.193	.188	.184	.177	.172	.169	.164	.161	.158	.154
45°	.239	.227	.218	.209	.202	.195	.188	.183	.178	.173	.168	.165	.161	.157	.154	.151

Ratios of W to W'.

TABLE B.

Table of coefficients of h for finding the Thickness of Standard Rectangular Walls, when the top of the bank slopes away at the Angle of Repose.

Angle of Repose	|1	1·1	1·2	1·3	1·4	1·5	1·6 K	1·7 K	1·8 K	1·9	2	2·1	2·2	2·3	2·4	2·8
30°	.500	.476	.456	.438	.423	.409	.395	.384	.372	.363	.353	.345	.337	.330	.323	.316
31°	.494	.471	.452	.434	.418	.404	.391	.380	.368	.359	.350	.342	.333	.326	.320	.313
32°	.489	.466	.447	.429	.414	.400	.387	.376	.365	.355	.346	.338	.330	.323	.316	.309
33°	.483	.461	.442	.424	.409	.395	.382	.371	.360	.351	.342	.334	.326	.319	.312	.306
34°	.478	.456	.437	.419	.404	.391	.378	.367	.356	.347	.338	.331	.322	.316	.309	.303
35°	.472	.450	.431	.414	.399	.386	.373	.362	.352	.343	.334	.327	.318	.312	.305	.299
36°	.467	.445	.426	.409	.395	.382	.369	.358	.348	.339	.330	.323	.315	.308	.302	.295
37°	.461	.439	.421	.404	.389	.377	.364	.354	.343	.334	.326	.319	.311	.304	.298	.291
38°	.455	.433	.415	.399	.384	.372	.359	.349	.339	.330	.321	.314	.306	.300	.294	.288
39°	.448	.427	.409	.393	.379	.367	.354	.344	.334	.325	.317	.310	.302	.296	.290	.284
40°	.442	.421	.404	.388	.374	.361	.349	.339	.329	.321	.312	.306	.298	.292	.286	.279
41°	.435	.415	.397	.382	.368	.356	.344	.334	.324	.316	.307	.301	.293	.287	.281	.275
42°	.429	.409	.391	.376	.363	.351	.339	.329	.319	.311	.303	.296	.289	.283	.277	.271
43°	.422	.402	.385	.370	.357	.345	.333	.324	.314	.306	.298	.292	.284	.278	.273	.267
44°	.415	.395	.379	.364	.351	.339	.328	.318	.309	.301	.293	.287	.280	.274	.268	.262
45°	.408	.389	.373	.358	.345	.334	.322	.313	.304	.296	.288	.282	.275	.269	.264	.258

NOTE II.

The following graphic method for determining the pressure of earth against a retaining wall, we take from " Engineering :"

Referring to Fig. 1, let us determine, first, the pressure exerted by the wedge, A, B, C, the angle, B, A, E, being greater than Φ, the limited angle of resistance of the wedge against A, B, which is also identical with the natural slope of the earth. The friction against the back of the wall is neglected. We have now 3 forces to deal with—namely, the weight of the triangle, A, B, C, acting vertically through its centre of gravity, and therefore passing through the point c, where A, $c = \frac{1}{3}$ A, B ; next, the resistance of the plane A, B, the direction of which is inclined at an angle, Φ, to the normal to A, B ; and, lastly, the thrust against the back of the wall acting horizontally through c, and cutting A, C in a point, g, where A $g =^1$ A, C.

Now, since the weight of the wedge, A
B C, is proportional to C B, the height of
the wall remaining constant, and the slope
varying, if we set off $c\ b = $ C B, and complete the triangle of forces, $a\ b$ will represent the thrust against the back of the
wall.

Let us now observe the effect produced
by altering the slope A B, to A D.

Construct the triangle of forces $d\ e\ f$, as
before, making $d\ f = $ C D, to represent

FIG. 1.

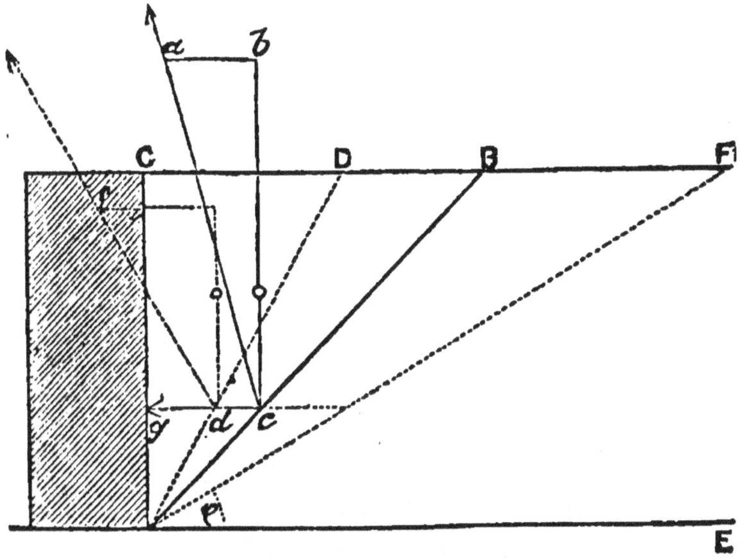

the weight of A C D. The angle, $e\ d\ f$, is
now greater than the angle $a\ c\ b$, by the

same amount that we have increased the slope of A B to A D; that is to say, the angle $e\,d\,f = a\,c\,b\,+$ angle D A B.

Also, the length of $b\,c$ has decreased to $f\,d$ in the ratio of C B to C D.

Supposing now we divide up the angle C A F into any number of equal angles by the radial lines $A_1\,A_2\,A_3$, etc. (see Fig. 2), and imagine the slope of A B to be altered

FIG. 2.

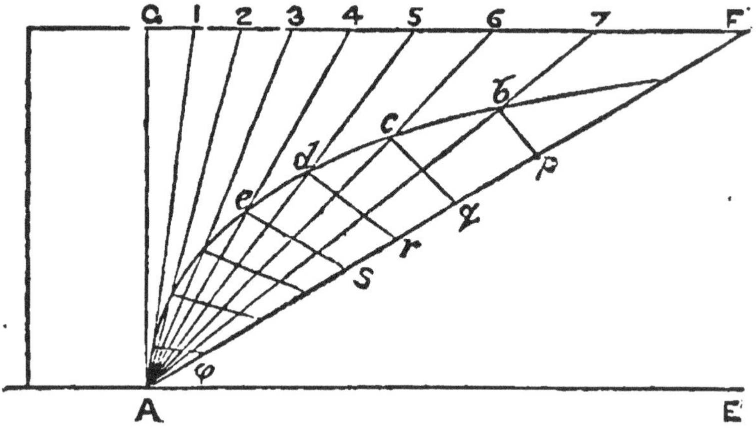

to each of these positions successively, we shall then for each alteration have a new triangle of forces; for instance, in moving from the position A_1 to A_2 the angle $b\,A\,p$, of the triangle of forces will increase to $c\,A$ q in the same ratio that the slope of the plane varies, and the side A b will decrease to A c in the same ratio that C_1 decreases to C_2.

We are thus enabled to make a diagram illustrating the successive changes by a curve, e d c b, A b, A c, A d, etc., being respectively equal to C_1, C_2, C_3, etc. The lines b p, c q, d r, e s, and c drawn at right angles to A b, A c, A d, etc., will now represent the thrusts against the back of the wall at the different slopes, and it will be observed on examining the diagram that the position which gives the greatest magnitude to the line representing the thrust is the slope A_4, which bisects the angle C A F.

VALUABLE
SCIENTIFIC BOOKS,

PUBLISHED BY

D. VAN NOSTRAND,

23 Murray Street and 27 Warren Street,

NEW YORK.

FRANCIS. Lowell Hydraulic Experiments, being a selection from Experiments on Hydraulic Motors, on the Flow of Water over Weirs, in Open Canals of Uniform Rectangular Section, and through submerged Orifices and diverging Tubes. Made at Lowell, Massachusetts. By James B. Francis, C. E. 2d edition, revised and enlarged, with many new experiments, and illustrated with twenty-three copperplate engravings. 1 vol. 4to, cloth......................$15 00.

ROEBLING (J. A.) Long and Short Span Railway Bridges. By John A. Roebling, C. E. Illustrated with large copperplate engravings of plans and views. Imperial folio, cloth............................. 25 00

CLARKE (T. C.) Description of the Iron Railway Bridge over the Mississippi River, at Quincy, Illinois. Thomas Curtis Clarke, Chief Engineer. Illustrated with 21 lithographed plans. 1 vol. 4to, cloth .... 7 50.

TUNNER (P.) A Treatise on Roll-Turning for the Manufacture of Iron. By Peter Tunner. Translated and adapted by John B. Pearse, of the Penn-

I

sylvania Steel Works, with numerous engravings wood cuts and folio atlas of plates.................$10 00

ISHERWOOD (B. F.) Engineering Precedents for Steam Machinery. Arranged in the most practical and useful manner for Engineers. By B. F. Isherwood, Civil Engineer, U. S. Navy. With Illustrations. Two volumes in one. 8vo, cloth.......... $2 50

BAUERMAN. Treatise on the Metallurgy of Iron, containing outlines of the History of Iron Manufacture, methods of Assay, and analysis of Iron Ores, processes of manufacture of Iron and Steel, etc., etc. By H. Bauerman. First American edition. Revised and enlarged, with an Appendix on the Martin Process for making Steel, from the report of Abram S. Hewitt. Illustrated with numerous wood engravings. 12mo, cloth.. 2 00

CAMPIN on the Construction of Iron Roofs. By Francis Campin. 8vo, with plates, cloth.......... 3 00

COLLINS. The Private Book of Useful Alloys and Memoranda for Goldsmiths, Jewellers, &c. By James E. Collins. 18mo, cloth.................. 75

CIPHER AND SECRET LETTER AND TELEGRAPHIC CODE, with Hogg's Improvements. The most perfect secret code ever invented or discovered. Impossible to read without the key. By C. S. Larrabee. 18mo, cloth. 1 00

COLBURN. The Gas Works of London. By Zerah Colburn, C. E. 1 vol. 12mo, boards.............. 60

CRAIG (B. F.) Weights and Measures. An account of the Decimal System, with Tables of Conversion for Commercial and Scientific Uses. By B. F. Craig, M.D. 1 vol. square 32mo, limp cloth............. 50

NUGENT. Treatise on Optics; or, Light and Sight, theoretically and practically treated; with the application to Fine Art and Industrial Pursuits. By E. Nugent. With one hundred and three illustrations. 12mo, cloth.................................... 2 00

GLYNN (J.) Treatise on the Power of Water, as applied to drive Flour Mills, and to give motion to Turbines and other Hydrostatic Engines. By Joseph

Glynn. Third edition, revised and enlarged, with numerous illustrations. 12mo, cloth. $1 00

HUMBER. A Handy Book for the Calculation of Strains in Girders and similar Structures, and their Strength, consisting of Formulæ and corresponding Diagrams, with numerous details for practical application. By William Humber. 12mo, fully illustrated, cloth. 2 50

GRUNER. The Manufacture of Steel. By M. L. Gruner. Translated from the French, by Lenox Smith, with an appendix on the Bessamer process in the United States, by the translator. Illustrated by Lithographed drawings and wood cuts. 8vo, cloth.. 3 50

AUCHINCLOSS. Link and Valve Motions Simplified. Illustrated with 37 wood-cuts, and 21 lithographic plates, together with a Travel Scale, and numerous useful Tables. By W. S. Auchincloss. 8vo, cloth.. 3 00

VAN BUREN. Investigations of Formulas, for the strength of the Iron parts of Steam Machinery. By J. D. Van Buren, Jr., C. E. Illustrated, 8vo, cloth. 2 00

JOYNSON. Designing and Construction of Machine Gearing. Illustrated, 8vo, cloth......... .. .. 2 00

GILLMORE. Coignet Beton and other Artificial Stone. By Q. A. Gillmore, Major U S. Corps Engineers. 9 plates, views, &c. 8vo, cloth.................... 2 50

SAELTZER. Treatise on Acoustics in connection with Ventilation. By Alexander Saeltzer, Architect. 12mo, cloth................................... 2 00

THE EARTH'S CRUST. A handy Outline of Geology. By David Page. Illustrated, 18mo, cloth.... 75

DICTIONARY of Manufactures, Mining, Machinery, and the Industrial Arts. By George Dodd. 12mo, . cloth.. 2 00

FRANCIS. On the Strength of Cast-Iron Pillars, with Tables for the use of Engineers, Architects, and Builders. By James B. Francis, Civil Engineer. 1 vol. 8vo, cloth................................ 2 50

GILLMORE (Gen. Q. A.) Treatise on Limes, Hydraulic Cements, and Mortars. Papers on Practical Engineering, U. S. Engineer Department, No. 9, containing Reports of numerous Experiments conducted in New York City, during the years 1858 to 1861, inclusive. By Q. A. Gillmore, Bvt. Maj -Gen., U. S. A., Major, Corps of Engineers. With numerous illustrations. 1 vol, 8vo, cloth.............. $4 00

HARRISON. The Mechanic's Tool Book, with Practical Rules and Suggestions for Use of Machinists, Iron Workers, and others. By W. B. Harrison, associate editor of the "American Artisan." Illustrated with 44 engravings. 12mo, cloth........... 1 50

HENRICI (Olaus). Skeleton Structures, especially in their application to the Building of Steel and Iron Bridges. By Olaus Henrici. With folding plates and diagrams. 1 vol. 8vo, cloth................... 3 00

HEWSON (Wm.) Principles and Practice of Embanking Lands from River Floods, as applied to the Levees of the Mississippi. By William Hewson, Civil Engineer. 1 vol. 8vo, cloth..................... 2 00

HOLLEY (A. L.) Railway Practice. American and European Railway Practice, in the economical Generation of Steam, including the Materials and Construction of Coal-burning Boilers, Combustion, the Variable Blast, Vaporization, Circulation, Superheating, Supplying and Heating Feed-water, etc., and the Adaptation of Wood and Coke-burning Engines to Coal-burning; and in Permanent Way, including Road-bed, Sleepers, Rails, Joint-fastenings, Street Railways, etc., etc. By Alexander L. Holley, B. P. With 77 lithographed plates. 1 vol. folio, cloth.... 12 00

KING (W. H.) Lessons and Practical Notes on Steam, the Steam Engine, Propellers, etc., etc., for Young Marine Engineers, Students, and others. By the late W. H. King, U. S. Navy. Revised by Chief Engineer J. W. King, U. S. Navy. Twelfth edition, enlarged. 8vo, cloth. 2 00

MINIFIE (Wm.) Mechanical Drawing. A Text-Book of Geometrical Drawing for the use of Mechanics and Schools, in which the Definitions and Rules of Geometry are familiarly explained; the Practical Problems are arranged, from the most simple to the more complex, and in their description technicalities are avoided as much as possible. With illustrations for Drawing Plans, Sections, and Elevations of Railways and Machinery; an Introduction to Isometrical Drawing, and an Essay on Linear Perspective and Shadows. Illustrated with over 200 diagrams engraved on steel. By Wm. Minifie, Architect. Seventh edition. With an Appendix on the Theory and Application of Colors. 1 vol. 8vo, cloth........... $4 00

"It is the best work on Drawing that we have ever seen, and is especially a text-book of Geometrical Drawing for the use of Mechanics and Schools. No young Mechanic, such as a Machinists, Engineer, Cabinet-maker, Millwright, or Carpenter, should be without it."—*Scientific American.*

———— Geometrical Drawing. Abridged from the octavo edition, for the use of Schools. Illustrated with 48 steel plates. Fifth edition. 1 vol. 12mo, cloth.... 2 00

STILLMAN (Paul.) Steam Engine Indicator, and the Improved Manometer Steam and Vacuum Gauges— their Utility and Application. By Paul Stillman. New edition. 1 vol. 12mo, flexible cloth.......... 1 00

SWEET (S. H.) Special Report on Coal; showing its Distribution, Classification, and cost delivered over different routes to various points in the State of New York, and the principal cities on the Atlantic Coast. By S. H. Sweet. With maps, 1 vol. 8vo, cloth..... 3 00

WALKER (W. H.) Screw Propulsion. Notes on Screw Propulsion: its Rise and History. By Capt. W. H. Walker, U. S. Navy. 1 vol. 8vo, cloth..... 75

WARD (J. H.) Steam for the Million. A popular Treatise on Steam and its Application to the Useful Arts, especially to Navigation. By J. H. Ward, Commander U. S. Navy. New and revised edition. 1 vol. 8vo, cloth................................. 1 00

WEISBACH (Julius). Principles of the Mechanics of Machinery and Engineering. By Dr. Julius Weisbach, of Freiburg. Translated from the last German edition. 1 Vol. I., 8vo, cloth.. .... 10 00

DIEDRICH. The Theory of Strains. a Compendium for the calculation and construction of Bridges, Roofs, and Cranes, with the application of Trigonometrical Notes, containing the most comprehensive information in regard to the Resulting strains for a permanent Load, as also for a combined (Permanent and Rolling) Load. In two sections, adadted to the requirements of the present time. By John Diedrich, C. E. Illustrated by numerous plates and diagrams. 8vo, cloth... 5 00

WILLIAMSON (R. S.) On the use of the Barometer on Surveys and Reconnoissances. Part I. Meteorology in its Connection with Hypsometry. Part II. Barometric Hypsometry. By R. S. Wiliamson, Bvt. Lieut.-Col. U. S. A., Major Corps of Engineers. With Illustrative Tables and Engravings. Paper No. 15, Professional Papers, Corps of Engineers. 1 vol. 4to, cloth..................................... 15 00

POOK (S. M.) Method of Comparing the Lines and Draughting Vessels Propelled by Sail or Steam. Including a chapter on Laying off on the Mould-Loft Floor. By Samuel M. Pook, Naval Constructor. 1 vol. 8vo, with illustrations, cloth........... 5 00

ALEXANDER (J. H.) Universal Dictionary of Weights and Measures, Ancient and Modern, reduced to the standards of the United States of America. By J. H. Alexander. New edition, enlarged. 1 vol. 8vo, cloth............................... 3 50

BROOKLYN WATER WORKS. Containing a Descriptive Account of the Construction of the Works, and also Reports on the Brooklyn, Hartford, Belleville and Cambridge Pumping Engines. With illustrations. 1 vol. folio, cloth......................... 20 00

RICHARDS' INDICATOR. A Treatise on the Richards Steam Engine Indicator, with an Appendix by F. W. Bacon, M. E. 18mo, flexible, cloth.......... 1 00

POPE. Modern Practice of the Electric Telegraph. A Hand Book for Electricians and operators. By Frank L. Pope. Eighth edition, revised and enlarged, and fully illustrated. 8vo, cloth.................... $2.00

"There is no other work of this kind in the English language that contains in so small a compass so much practical information in the application of galvanic electricity to telegraphy. It should be in the hands of every one interested in telegraphy, or the use of Batteries for other purposes."

MORSE. Examination of the Telegraphic Apparatus and the Processes in Telegraphy. By Samuel F. Morse, LL.D., U. S. Commissioner Paris Universal Exposition, 1867. Illustrated, 8vo, cloth.......... $2 00

SABINE. History and Progress of the Electric Telegraph, with descriptions of some of the apparatus. By Robert Sabine, C. E. Second edition, with additions, 12mo, cloth.............................. 1 25

CULLEY. A Hand-Book of Practical Telegraphy. By R. S. Culley, Engineer to the Electric and International Telegraph Company. Fourth edition, revised and enlarged. Illustrated 8vo, cloth.............. 5 00

BENET. Electro-Ballistic Machines, and the Schultz Chronoscope. By Lieut.-Col. S. V Benet, Captain of Ordnance, U. S. Army. Illustrated, second edition, 4to, cloth.................................. 3 00

MICHAELIS. The Le Poulenge Chronograph, with three Lithograph folding plates of illustrations. By Brevet Captain O. E. Michaelis, First Lieutenant Ordnance Corps, U. S. Army, 4to, cloth.......... 3 00

ENGINEERING FACTS AND FIGURES An Annual Register of Progress in Mechanical Engineering and Construction. for the years 1863, 64, 65, 66, 67, 68. Fully illustrated, 6 vols. 18mo, cloth, $2.50 per vol., each volume sold separately.............

HAMILTON. Useful Information for Railway Men. Compiled by W. G. Hamilton, Engineer. Fifth edition, revised and enlarged, 562 pages Pocket form. Morocco, gilt................................... 2 00

STUART. The Civil and Military Engineers of America. By Gen. C. B. Stuart. With 9 finely executed portraits of eminent engineers, and illustrated by engravings of some of the most important works constructed in America. 8vo, cloth.................... $5 00

STONEY. The Theory of Strains in Girders and similar structures, with observations on the application of Theory to Practice, and Tables of Strength and other properties of Materials. By Bindon B. Stoney, B. A. New and revised edition, enlarged, with numerous engravings on wood, by Oldham. Royal 8vo, 664 pages. Complete in one volume. 8vo, cloth....... 15 00

SHREVE. A Treatise on the Strength of Bridges and Roofs. Comprising the determination of Algebraic formulas for strains in Horizontal, Inclined or Rafter, Triangular, Bowstring, Lenticular and other Trusses, from fixed and moving loads, with practical applications and examples, for the use of Students and Engineers. By Samuel H. Shreve, A. M., Civil Engineer. 87 wood cut illustrations. 8vo, cloth.............. 5 00

MERRILL. Iron Truss Bridges for Railroads. The method of calculating strains in Trusses, with a careful comparison of the most prominent Trusses, in reference to economy in combination, etc., etc. By Brevet Col. William E. Merrill, U S. A., Major Corps of Engineers, with nine lithographed plates of Illustrations. 4to, cloth...................... 5 00

WHIPPLE. An Elementary and Practical Treatise on Bridge Building. An enlarged and improved edition of the author's original work. By S. Whipple, C. E., inventor of the Whipple Bridges, &c. Illustrated 8vo, cloth...................................... 4 00

THE KANSAS CITY BRIDGE. With an account of the Regimen of the Missouri River, and a description of the methods used for Founding in that River. By O Chanute, Chief Engineer, and George Morrison, Assistant Engineer. Illustrated with five lithographic views and twelve plates of plans. 4to, cloth, 6 00

8

MAC CORD. A Practical Treatise on the Slide Valve by Eccentrics, examining by methods the action of the Eccentric upon the Slide Valve, and explaining the Practical processes of laying out the movements, adapting the valve for its various duties in the steam engine. For the use of Engineers, Draughtsmen, Machinists, and Students of Valve Motions in general. By C. W. Mac Cord, A. M., Professor of Mechanical Drawing, Stevens' Institute of Technology, Hoboken, N. J. Illustrated by 8 full page copper-plates. 4to, cloth............................... $4 00

KIRKWOOD. Report on the Filtration of River Waters, for the supply of cities, as practised in Europe, made to the Board of Water Commissioners of the City of St. Louis. By James P. Kirkwood. Illustrated by 30 double plate engravings. 4to, cloth, 15 00

PLATTNER. Manual of Qualitative and Quantitative Analysis with the Blow Pipe. From the last German edition, revised and enlarged. By Prof. Th. Richter, of the Royal Saxon Mining Academy. Translated by Prof. H. B. Cornwall, Assistant in the Columbia School of Mines, New York assisted by John H. Caswell. Illustrated with 87 wood cuts, and one lithographic plate. Second edition, revised, 560 pages, 8vo, cloth................................ 7 50

PLYMPTON. The Blow Pipe. A system of Instruction in its practical use being a graduated course of analysis for the use of students, and all those engaged in the examination of metallic combinations Second edition, with an appendix and a copious index. By Prof. Geo W. Plympton, of the Polytechnic Institute, Brooklyn, N. Y. 12mo, cloth............... 2 00

PYNCHON. Introduction to Chemical Physics, designed for the use of Academies, Colleges and High Schools. Illustrated with numerous engravings, and containing copious experiments with directions for preparing them. By Thomas Ruggles Pynchon, M. A., Professor of Chemistry and the Natural Sciences, Trinity College, Hartford New edition, revised and enlarged and illustrated by 269 illustrations on wood. Crown, 8vo. cloth.................... 3 00

9

ELIOT AND STORER. A compendious Manual of Qualitative Chemical Analysis. By Charles W. Eliot and Frank H. Storer. Revised with the Co-operation of the authors. By William R. Nichols, Professor of Chemistry in the Massachusetts Institute of Technology Illustrated, 12mo, cloth....... $1 50

RAMMELSBERG. Guide to a course of Quantitative Chemical Analysis. especially of Minerals and Furnace Products. Illustrated by Examples By C. F. Rammelsberg. Translated by J. Towler, M. D. 8vo, cloth.................................... 2 25

EGLESTON. Lectures on Descriptive Mineralogy, delivered at the School of Mines. Columbia College. By Professor T. Egleston. Illustrated by 34 Lithographic Plates. 8vo, cloth...................... 4 50

MITCHELL. A Manual of Practical Assaying. By John Mitchell. Third edition. Edited by William Crookes, F. R. S. 8vo, cloth................... 10 00

WATT'S Dictionary of Chemistry. New and Revised edition complete in 6 vols. 8vo cloth, $62.00 Supplementary volume sold separately. Price, cloth... 9 00

RANDALL. Quartz Operators Hand-Book. By P. M. Randall. New edition, revised and enlarged, fully illustrated. 12mo, cloth...................... 2 00

SILVERSMITH. A Practical Hand-Book for Miners, Metallurgists, and Assayers, comprising the most recent improvements in the disintegration. amalgamation, smelting, and parting of the precious ores, with a comprehensive Digest of the Mining Laws. Greatly augmented, revised and corrected. By Julius Silversmith. Fourth edition. Profusely illustrated. 12mo, cloth... 3 00

THE USEFUL METALS AND THEIR ALLOYS, including Mining Ventilation, Mining Jurisprudence, and Metallurgic Chemistry employed in the conversion of Iron, Copper, Tin, Zinc, Antimony and Lead ores, with their applications to the Industrial Arts. By Scoffren, Truan, Clay, Oxland, Fairbairn, and others. Fifth edition, half calf.................... 3 75

JOYNSON. The Metals used in construction, Iron, Steel, Bessemer Metal, etc., etc. By F. H. Joynson. Illustrated, 12mo, cloth............................ $0 75

VON COTTA. Treatise on Ore Deposits. By Bernhard Von Cotta, Professor of Geology in the Royal School of Mines, Freidberg, Saxony. Translated from the second German edition, by Frederick Prime, Jr., Mining Engineer, and revised by the author, with numerous illustrations. 8vo, cloth....... 4 00

URE. Dictionary of Arts, Manufactures and Mines.' By Andrew Ure, M.D. Sixth edition, edited by Robert Hunt, F. R. S., greatly enlarged and re-written. London, 1872. 3 vols. 8vo, cloth, $25.00. Half Russia... 37 50

BELL. Chemical Phenomena of Iron Smelting. An experimental and practical examination of the circumstances which determine the capacity of the Blast Furnace, The Temperature of the air, and the proper condition of the Materials to be operated upon. By I. Lowthian Bell. 8vo, cloth........... 6 00

ROGERS. The Geology of Pennsylvania. A Government survey, with a general view of the Geology of the United States, Essays on the Coal Formation and its Fossils, and a description of the Coal Fields of North America and Great Britain. By Henry Darwin Rogers, late State Geologist of Pennsylvania, Splendidly illustrated with Plates and Engravings in the text. 3 vols., 4to, cloth, with Portfolio of Maps. 30 00

BURGH. Modern Marine Engineering, applied to Paddle and Screw Propulsion. Consisting of 36 colored plates, 259 Practical Wood Cut Illustrations, and 403 pages of descriptive matter, the whole being an exposition of the present practice of James Watt & Co., J. & G. Rennie, R. Napier & Sons, and other celebrated firms, by N. P. Burgh, Engineer, thick 4to, vol., cloth, $25.00; half mor........ 30 00

BARTOL. Treatise on the Marine Boilers of the United States, By B. H. Bartol. Illustrated, 8vo, cloth... 1 50

BOURNE. Treatise on the Steam Engine in its various applications to Mines, Mills, Steam Navigation, Railways, and Agriculture, with the theoretical investigations respecting the Motive Power of Heat, and the proper proportions of steam engines. Elaborate tables of the right dimensions of every part, and Practical Instructions for the manufacture and management of every species of Engine in actual use. By John Bourne, being the ninth edition of "A Treatise on the Steam Engine," by the "Artizan Club." Illustrated by 38 plates and 546 wood cuts. 4to, cloth..$15 00

STUART. The Naval Dry Docks of the United States. By Charles B. Stuart late Engineer-in-Chief of the U. S. Navy. Illustrated with 24 engravings on steel. Fourth edition, cloth.................... 6 00

EADS. System of Naval Defences. By James B. Eads, C. E., with 10 illustrations, 4to, cloth........ 5 00

FOSTER. Submarine Blasting in Boston Harbor, Massachusetts. Removal of Tower and Corwin Rocks. By J. G. Foster, Lieut.-Col. of Engineers, U. S. Army. Illustrated with seven plates, 4to, cloth.. 3 50

BARNES Submarine Warfare, offensive and defensive, including a discussion of the offensive Torpedo System, its effects upon Iron Clad Ship Systems and influence upon future naval wars. By Lieut.-Commander J. S. Barnes, U. S. N., with twenty lithographic plates and many wood cuts. 8vo, cloth..... 5 00

HOLLEY. A Treatise on Ordnance and Armor, embracing descriptions, discussions, and professional opinions concerning the materials, fabrication, requirements, capabilities, and endurance of European and American Guns, for Naval, Sea Coast, and Iron Clad Warfare, and their Rifling, Projectiles, and Breech-Loading ; also, results of experiments against armor, from official records, with an appendix referring to Gun Cotton, Hooped Guns, etc., etc. By Alexander L. Holley, B. P., 948 pages, 493 engravings, and 147 Tables of Results, etc., 8vo, half roan. 10 00

SIMMS. A Treatise on the Principles and Practice of Levelling, showing its application to purposes of Railway Engineering and the Construction of Roads, &c. By Frederick W. Simms, C. E. From the 5th London edition, revised and corrected, with the addition of Mr. Laws's Practical Examples for setting out Railway Curves. Illustrated with three Lithographic plates and numerous wood cuts. 8vo, cloth. $2 50

BURT. Key to the Solar Compass, and Surveyor's Companion; comprising all the rules necessary for use in the field; also description of the Linear Surveys and Public Land System of the United States, Notes on the Barometer, suggestions for an outfit for a survey of four months, etc. By W. A. Burt, U. S. Deputy Surveyor. Second edition. Pocket book form, tuck... 2o 5o

THE PLANE TABLE. Its uses in Topographical Surveying, from the Papers of the U. S. Coast Survey. Illustrated, 8vo, cloth...................... 2 oo

"This work gives a description of the Plane Table, employed at the U. S. Coast Survey office, and the manner of using it."

JEFFER'S. Nautical Surveying. By W. N. Jeffers, Captain U. S. Navy. Illustrated with 9 copperplates and 31 wood cut illustrations. 8vo, cloth.......... 5 oo

CHAUVENET. New method of correcting Lunar Distances, and improved method of Finding the error and rate of a chronometer, by equal altitudes. By W. Chauvenet, LL.D. 8vo, cloth............... 2 oo

BRUNNOW. Spherical Astronomy. By F. Brunnow, Ph. Dr. Translated by the author from the second German edition. 8vo, cloth..................... 6 5o

PEIRCE. System of Analytic Mechanics. By Benjamin Peirce. 4to, cloth........................ 10 oo

COFFIN. Navigation and Nautical Astronomy. Prepared for the use of the U. S. Naval Academy. By Prof. J. H. C. Coffin. Fifth edition. 52 wood cut illustrations. 12mo, cloth 3 5o

13

D. VAN NOSTRAND'S PUBLICATIONS.

MYER. Manual of Signals. for the use of Signal officers in the Field, and for Military and Naval Students, Military Schools, etc. A new edition enlarged and illustrated. By Brig. General Albert J. Myer, Chief Signal Officer of the army, Colonel of the Signal Corps during the War of the Rebellion. 12mo, 48 plates, full Roan.................................. $5

WILLIAMSON. Practical Tables in Meteorology and Hypsometry, in connection with the use of the Barometer. By Col. R. S. Williamson, U. S. A. 4to, cloth.. 2

THE YOUNG MECHANIC. Containing directions for the use of all kinds of tools, and for the construction of Steam Engines and Mechanical Models, including the Art of Turning in Wood and Metal. By the author "The Lathe and its Uses," etc. From the English edition with corrections. Illustrated, 12mo, cloth...................................... 1

PICKERT AND METCALF. The Art of Graining. How Acquired and How Produced, with description of colors, and their application. By Charles Pickert and Abraham Metcalf. Beautifully illustrated with 42 tinted plates of the various woods used in interior finishing. Tinted paper, 4to, cloth.............. 10

HUNT. Designs for the Gateways of the Southern Entrances to the Central Park. By Richard M. Hunt. With a description of the designs. 4to. cloth...... 5

LAZELLE. One Law in Nature. By Capt. H. M. Lazelle, U. S. A. A new Corpuscular Theory, comprehending Unity of Force, Identity of Matter, and its Multiple Atom Constitution, applied to the Physical Affections or Modes of Energy. 12mo, cloth... 1

PETERS. Notes on the Origin, Nature, Prevention, and Treatment of Asiatic Cholera. By John C. Peters, M. D. Second edition, with an Appendix. 12mo, cloth....................................... 1

15

BOYNTON. History of West Point, its Military Importance during the American Revolution, and the Origin and History of the U. S. Military Academy. By Bvt. Major C. E. Boynton, A.M., Adjutant of the Military Academy. Second edition, 416 pp. 8vo, printed on tinted paper, beautifully illustrated with 36 maps and fine engravings, chiefly from photographs taken on the spot by the author. Extra cloth... $3 50

WOOD. West Point Scrap Book, being a collection of Legends, Stories, Songs, etc., of the U S. Military Academy. By Lieut. O E. Wood, U. S. A. Illustrated by 69 engravings and a copperplate map. Beautifully printed on tinted paper. 8vo, cloth..... 5 00

WEST POINT LIFE. A Poem read before the Dialectic Society of the United States Military Academy. Illustrated with Pen-and-Ink Sketches. By a Cadet. To which is added the song, "Benny Havens, oh!" oblong 8vo, 21 full page illustrations, cloth......... 2 50

GUIDE TO WEST POINT and the U. S. Military Academy, with maps and engravings, 18mo, blue cloth, flexible............................... .... 1 00

HENRY. Military Record of Civilian Appointments in the United States Army By Guy V. Henry, Brevet Colonel and Captain First United States Artillery, Late Colonel and Brevet Brigadier General, United States Volunteers. Vol. 1 now ready. Vol. 2 in press. 8vo, per volume, cloth.................... 5 00

HAMERSLY. Records of Living Officers of the U. S. Navy and Marine Corps. Compiled from official sources. By Lewis R. Hamersly, late Lieutenant U. S. Marine Corps. Revised edition, 8vo, cloth... 5 00

MOORE. Portrait Gallery of the War. Civil, Military and Naval. A Biographical record, edited by Frank Moore. 60 fine portraits on steel. Royal 8vo, cloth.. 6 00

www.ingramcontent.com/pod-product-compliance
Lightning Source LLC
Chambersburg PA
CBHW021524090426
42739CB00007B/766